In 1724, Stephen Hales, Vicar of Teddington and a pioneer of experimental plant physiology, performed numerous experiments to show 'the quantities imbibed and perspired by Plants and Trees'. Let him speak for himself:

> I took a garden-pot [Fig. 1] with a large Sun-flower, 3 feet + ½ high, which was purposely planted in it when young. I covered the pot with a plate of thin milled lead, and cemented all the joints fast, so as no vapour could pass, but only air, through a small glass tube.... Through [another] tube I watered the plant, and then stopped it up with a cork.... I weighed this pot and plant morning and evening, from July 3 to August 8.

After allowing for water evaporated through the unglazed walls of the pot he 'found the greatest perspiration of twelve hours in a very warm dry day, to be one pound fourteen ounces; the middle rate of perspiration one pound four ounces.'

At the end of the experiment the sunflower plant

FIG. 1. Drawing of a sunflower plant in a sealed pot which was weighed daily in 1724 by Stephen Hales to follow its transpiration; from his book *Vegetable staticks*.

was cut down and weighed. Its weight was 3 lb (1·36 kg); of which 2 lb 4 oz (1·2 kg) proved to be water. The loss of water by 'perspiration' on a warm day was therefore similar in amount to the water content of the plant and during the fifteen days of the experiment the plant must have 'perspired', or as we would say, transpired, nearly ten times its own weight in water.

In further experiments Hales investigated the ability of the soil to store water sufficient to meet the demands of plants in rainless periods and he measured gains of water by dew and loss by evaporation from the soil. An experiment on the uptake of water by a hop plant led him to multiply the daily uptake by the number of hop plants per acre (0·4 ha) and by the number of days in the growing season and so calculate the total transpiration of the crop as 0·9 in. (23·8 mm) of water spread over the acre (0·4 ha). He considered that the annual loss by evaporation from the soil was 6·2 in (157 mm) depth of water and that, from an annual rainfall of 22 in (558 mm) there would therefore remain about 15 in (380 mm) to feed springs. (These figures should not be taken to be quantitatively reliable. Transpiration in particular appears to have been underestimated.)

Finally he speculated about the physiological significance of the process.

> This quantity of moisture in a kindly state of the air is daily carried off, in a sufficient quantity, to keep the hops in a healthy state; but in a rainy moist state of air . . . too much moisture hovers about the hops, so as to hinder in a good measure the kindly perspiration of the leaves, whereby the stagnating sap corrupts, and breeds a moldy fen, which often spoils vast quantities of flourishing hop grounds.

Hales's speculations about the physiological significance of transpiration were mistaken, and it was probably the humid air around the plants rather than the stagnation of the sap within them that was responsible for the 'mouldy fen', probably the hop mould, *Sphaerotheca humuli*. But his scientific attitudes were essentially modern and he asked, and designed experiments to answer, almost all the questions which interest us today. What are the quantities, and what factors control the rates, of transpiration? How far can the demand for water be met by the soil? What is the significance of transpiration to the plant? What part does it play in the water balance of the earth's surface? Before re-examining these questions, we must make a new beginning.

Definition of transpiration

Transpiration is the process in which water is evaporated from plants. Since, as Hales showed, a plant may readily transpire the equivalent of its own water content in a day and yet its actual water content shows only relatively small daily fluctuations, it is clear that the uptake of water by the roots must fairly closely balance the loss by transpiration. It is convenient to distinguish three pathways or processes: uptake by the roots; movement of liquid water through the plant, sometimes called the transpiration stream; and evaporation and movement of water vapour away from the plant. The term transpiration is correctly applied only to the last of these three.

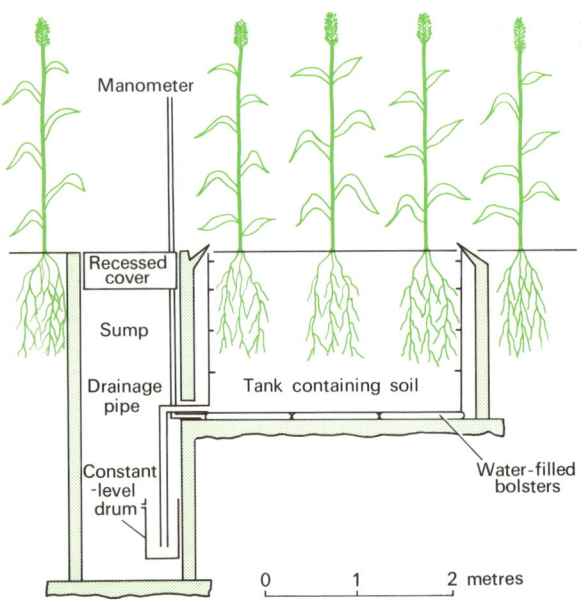

FIG. 2. Elevation of a lysimeter. It consists of a large water-tight tank filled with soil and of a depth adequate to allow the natural development of the roots of the crop. The tank is supported on a series of water-filled bolsters and placed in a pit with adjacent sump to receive and measure drainage. Changes in mass of the tank, due to water application, drainage, or transpiration, result in changes in pressure in the water-filled bolsters and hence in changes in the height of water in an open-ended manometer which provides hydrostatic balance for the system. (Adapted by permission from J. A. Forsgate *et al.* (1965), *Agric. Meteorol.* **2**.)

Measuring transpiration

Before investigating the processes that control transpiration we need methods of measuring it. Initial experimentation showed that transpiration rate is very sensitive to environmental factors such as temperature, air humidity, and wind. If we want to investigate the transpiration of a plant in nature it follows that any method of measurement that alters the environment is unsatisfactory. On the other hand, for the derivation of laws relating transpiration rate to environmental factors, it does not matter if these factors are artificially altered or controlled, provided they are accurately measured. Perhaps the two most obvious methods of measuring transpiration are either to follow the loss in weight of a plant/soil system, with the surface of the soil covered (like Hales's potted sunflower), or to measure the water vapour evolved from the plant or some of its leaves. Although a plant in a pot is hardly in a natural environment, the weighing technique can be extended to field conditions if a crop, or plant community, or single plant in a community, is grown in a container of soil of sufficient depth to allow the development of a normal root system, and the container is sunk with its rim level with the surrounding soil. Such a container is usually called a lysimeter (Fig. 2), although strictly this means a device for studying drainage from the soil. If the lysimeter can be weighed it is closely comparable with Hales's flower pot. More crudely, evaporation from the lysimeter can be estimated from the difference between rainfall and drainage. Note however that unless the surface of the soil is covered, and this is often not practicable, either method of using the lysimeter gives the sum of transpiration and evaporation from the soil. In dense plant communities the latter may be less than 10 per cent of the former but it may be considerably greater in open communities. It has sometimes been possible to construct a lysimeter around an undisturbed volume of soil and the plants that it supports.

Direct measurement of the water vapour evolved from a leaf or plant usually requires enclosure in a chamber and this almost inevitably alters the conditions in which transpiration occurs. However, the technique is well suited to investigating the responses of transpiration to experimentally induced alterations in air temperature, humidity, or windspeed. The water vapour evolved is fairly readily calculated from a comparison of the relative humidity and temperature of the air entering the cham-

ber and the air leaving it. (Students with a knowledge of physics might work this out in detail.)

An elegant method that can be used in natural conditions, theoretically and technically difficult but causing a minimum of disturbance to the transpiring plants, is to measure the humidities and wind-speeds at two heights above vegetation. From a knowledge of the physics of air mixing it is possible to calculate the rate of evolution of water vapour from the vegetation (see Fig. 3).

These by no means exhaust the methods, direct or indirect, that have been used to measure transpiration. The potometer (literally, 'drink measurer') is widely used in elementary studies to measure the rate of uptake of water by a detached leaf or shoot. It estimates transpiration of the detached part only if one assumes that water uptake and transpiration are equal, which is not necessarily true if environmental conditions are frequently changed. The reader should be able to devise further methods himself, and to criticise both his own and those described above. The potometer, in particular, requires more critical consideration than is usually given to it.

The conclusion can now be drawn that although it is fairly simple to measure the transpiration of a laboratory plant or part of a plant, measuring transpiration in nature may be practically laborious and expensive. Consequently the transpiration rates of natural vegetation and crops are often approximately estimated from formulae. These formulae are satisfactory only in so far as they are soundly based in theory, which itself is derived from, and tested by, experiments. The first experiments were necessarily with greenhouse and laboratory plants and even non-living systems, and they led to experiments in crops, grassland, and forest which employ the techniques that have been outlined. An account of the theoretical analysis of transpiration will be given, but first some aspects of the structure of the transpiring plant must be examined.

Fig. 3. A mast 110 ft high in pine forest at Thetford Chase, Suffolk, with instruments for studying the vertical flux of water vapour and hence the transpiration of the forest. (By courtesy of the Institute of Hydrology, Natural Environment Research Council.)

The transpiring leaf

Fig. 4 shows a vertical section through part of a spinach leaf. The leaf is enclosed by a tissue called the epidermis, on the outer surface of which is the waxy cuticle. The epidermis contains numerous stomata, each consisting of a pore surrounded by two guard cells (Fig. 5), whose movements are able to close or open the pore. Stomatal pores are char-

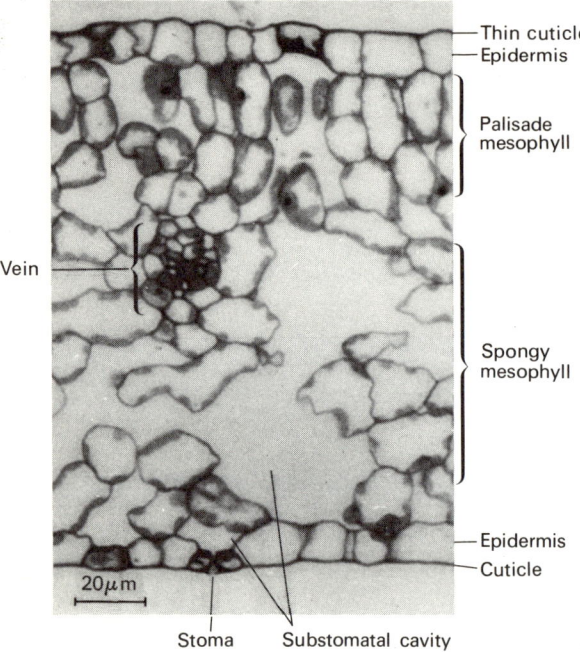

Vein

Thin cuticle
Epidermis

Palisade mesophyll

Spongy mesophyll

Epidermis
Cuticle

Stoma   Substomatal cavity

20 μm

FIG. 4. Vertical section through part of a spinach leaf. A stoma and guard cells are visible in the lower epidermis. There appears to be another stoma in the lower epidermis and two in the upper surface, just outside the plane of the section. (Photo by A. J. Collinge and A. D. Greenwood.)

FIG. 6. Demonstration of the boundary layer. A paper replica of a leaf is supported on a wire frame in a chamber through which air is blown from the right-hand side. A drop of smoke-forming liquid has been placed on the upper surface, right-hand side, of the leaf replica and has accumulated in the boundary layer, where indications of laminar flow can be seen. There is of course a boundary layer on the lower surface also, but the smoke is not accessible to it. (From D. J. Avery (1966), *J. exp. Bot.* **17.**)

acteristically between 10 and 30 μm long, according to species of plant, and 5–10 μm in diameter when fully open. Their densities are commonly between 50 and 200 per mm² leaf surface and the corresponding area of open pores is between 0·5 and 1·5 per cent of the total leaf area. Each pore leads into a substomatal cavity bounded by mesophyll cells, and smaller spaces between the cells are continuous with the substomatal cavities. All the intercellular spaces of the mesophyll, including the substomatal cavities, contain air.

If transpiration is measured in the light, with open stomata, and in the dark (which induces stomatal closure), and other conditions are kept constant, it is usually found that stomatal closure strongly reduces transpiration, although it may not entirely prevent it. (The experiment is not as simple to control as it sounds since light bright enough to cause stomatal opening tends to raise temperature.) Since transpiration occurs from surfaces without stomata, e.g. the upper surfaces of many leaves, as well as from surfaces with closed stomata, it is evident that some transpiration occurs by diffusion through the cuticle. However, cuticular transpiration is low in most plants (see, for example, Fig. 13), which leads to the conclusion that most of the water transpired must be evaporated from the internal surfaces of leaves (the surfaces of the mesophyll cells) and diffuse in the gaseous state through the substomatal cavities and the stomatal pores.

When the vapour molecules reach the outer ends of the pores they are still not at the end of the diffusion pathway for various kinds of evidence reveal the existence of a so-called boundary layer, external to the leaf, across which molecules must diffuse. Within the boundary layer the air is either still, or it shows what is described as laminar flow, i.e. layers of air move over one another parallel to the leaf surface and with increasing velocity at greater distance from the leaf, but with negligible mixing between adjacent layers. Outside the boundary layer air movement is turbulent, and vapour molecules are rapidly incorporated into the general body of the air. Avery has visually demonstrated

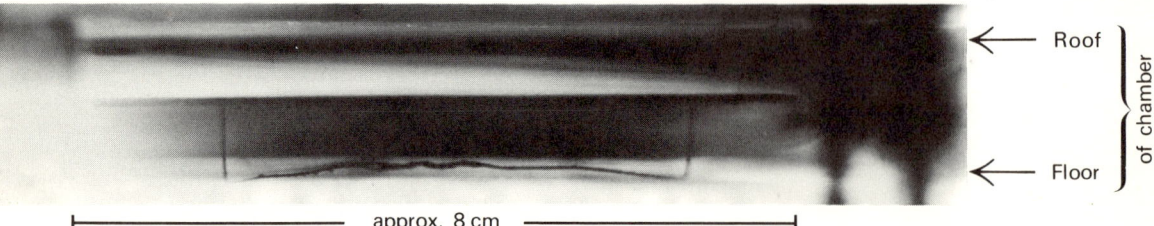

Roof

Floor

of chamber

approx. 8 cm

6

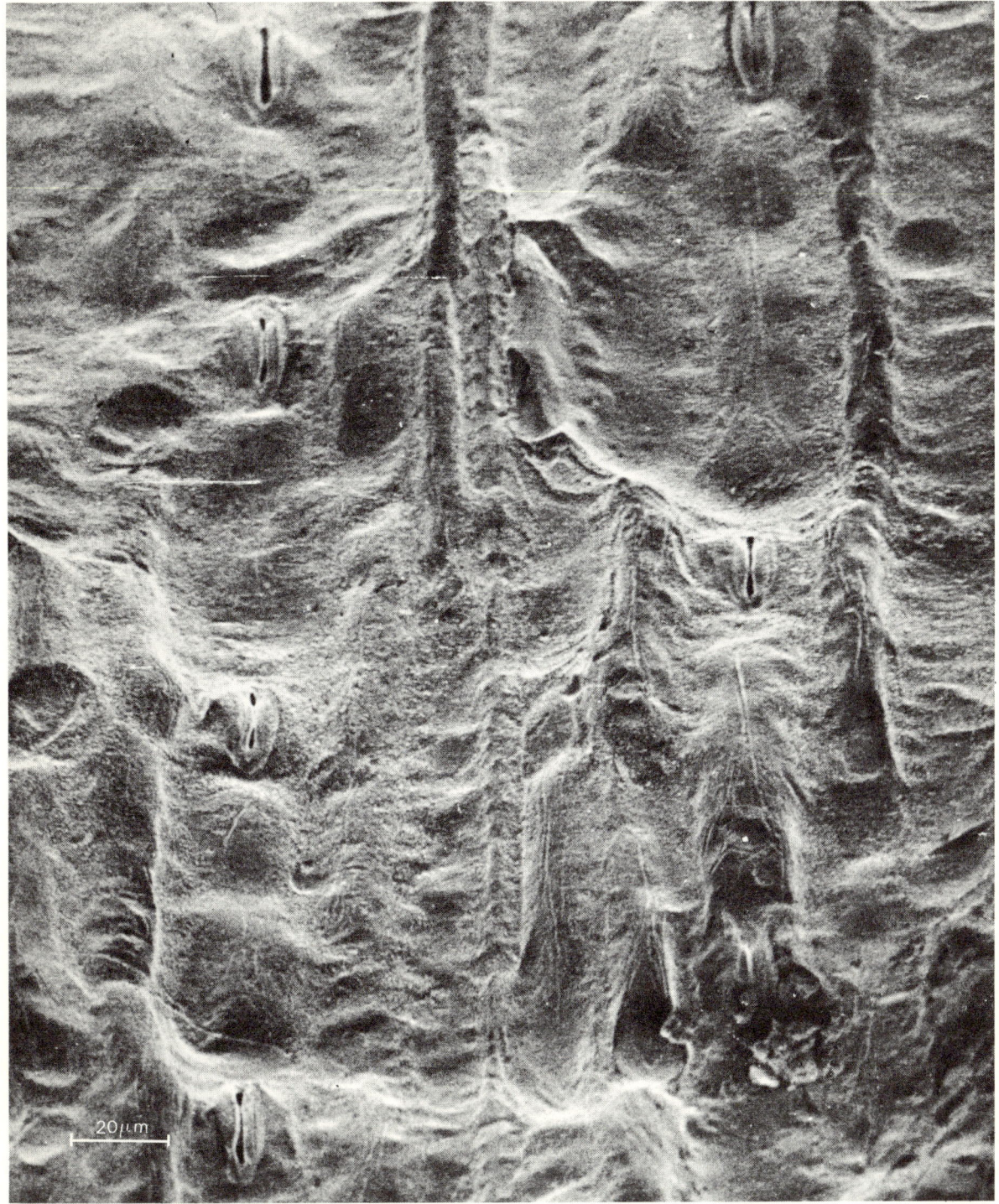

FIG. 5. Two rows of stomata seen in surface view of a maize leaf by stereoscan electron microscopy. The pores are elongated and are enclosed by two guard cells (sausage-shaped in surface view). (Photo by A. D. Greenwood and the Cambridge Instrument Co.)

the boundary layer by placing a drop of the smoke-forming liquid, titanium tetrachloride, on the leading edge of a leaf in a chamber through which air was blown at different speeds. His photograph (Fig. 6) shows the smoke accumulated in the boundary layer, but blown away from the rest of the chamber. By observing from the side with a travelling microscope he was able to measure the thickness of the boundary layer and demonstrate an inverse relation between thickness and wind-speed (Fig. 7). The thickness of the boundary layer also increases in proportion to the mean diameter of the leaf.

The existence of turbulence even in the relatively still air of a room can be observed in the swirling movements superimposed on the general rise (due to its warmth) of smoke, and the rapid disappearance of a limited amount of smoke illustrates the efficiency of turbulent mixing. By contrast gaseous diffusion through still air is a slow process, but even this is about 10 000 times as fast as diffusion of gases through water.

The stomata and air spaces in the mesophyll therefore constitute a much more rapid route for the inward diffusion of carbon dioxide for photosynthesis than would exist if the carbon dioxide had to diffuse from cell to cell in the liquid phase; and the provision of this relatively free path for the inward diffusion of carbon dioxide leads inevitably to an outward diffusion of water vapour unless the external air is saturated.

Having established that transpiration occurs mainly through the stomata our next experimental observation is that the rate at which water is transpired from a leaf with open stomata is commonly between 25 and 75 per cent of the rate of evaporation from a moist filter paper or a dish of water, of the same size and shape as the leaf, although only about 1 per cent of the leaf area is perforated and the cuticle is relatively impermeable. The explanation of this was discovered by investigations of the diffusion of gases through perforated septa made by Brown and Escombe in 1900. Their experiments were mainly on the diffusion of carbon dioxide, but are equally applicable to water vapour.

Some experiments on diffusion

Consider a tube (Fig. 8(a)) of length L and diameter D, containing concentrated sodium hydroxide solution at its lower end. The concentration of carbon dioxide in the air of the laboratory is C_a, at the mouth of the tube C_m, and at the surface of the solution C_s, g/cm³. C_s is effectively zero, and if the air is circulating over the mouth of the tube, C_m is effectively equal to C_a, but it will be convenient in further discussion to maintain these separate symbols. Carbon dioxide will diffuse from the external air down the concentration gradient created in the tube by absorption at the lower end, and it can be shown that the quantity Q of carbon dioxide diffusing and absorbed in time t, is proportional to the difference in concentration between the two ends of the tube, to the area A of the tube, and to the diffusion coefficient k' of carbon dioxide molecules in air; and is inversely proportional to the length of the tube, i.e.

$$Q = \frac{k'\,(C_m - C_s)\,A\,t}{L}\ \mathrm{g\,CO_2}$$

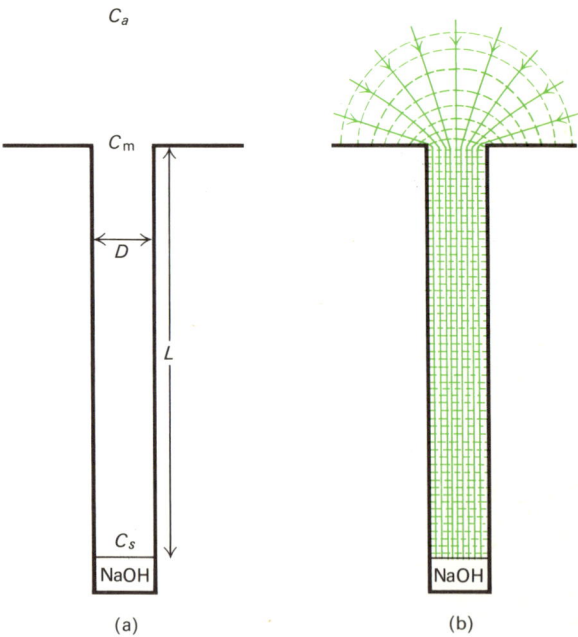

FIG. 8. Diffusion of carbon dioxide to absorbent sodium hydroxide at the bottom of a tube. The tube has a broad rim around its mouth. (a) Shows symbols for the dimensions of the tube and concentrations of carbon dioxide. (b) Shows direction of diffusion flux (continuous blue lines) and planes of uniform concentration (dotted blue lines).

More conveniently we define a *diffusive flux F* as the *rate* of diffusion across unit cross sectional area of the tube, i.e. $F = \dfrac{Q}{At}$ whence,

$$F = \frac{k'(C_m - C_s)}{L} \, g \, CO_2/cm^2/s \qquad (1)$$

We may regard the tube as offering a resistance to diffusive flux which is proportional to the length of still air, L cm.

Brown and Escombe found that if the tube was in still air there was also a resistance to diffusion up to the mouth of the tube, which was inversely proportional to its diameter; that is to say that the rate of diffusion *to unit area* of a circular orifice is more rapid for orifices of smaller diameter. In symbols:

$$F = \frac{k'(C_a - C_m)}{(\pi/8)D} \, g/cm^2/s \qquad (2a)$$

$$\text{or} \quad F = \frac{k'(C_a - C_m)}{0.392\,D} \qquad (2b)$$

They showed this by experimenting with dishes of different diameter (Fig. 9) filled to the brim with sodium hydroxide, so that $C_m = C_s = 0$, and they explained the result by the following considerations, theoretically derived by Stefan in 1881. The concentration of carbon dioxide increases outwards away from the sodium hydroxide until at theoretically infinite distance, but practically at 5–6 times the diameter of the dish, the normal concentration of air is found. The concentric, somewhat flattened semicircles shown in Fig. 9 represent concentric shells each of a given concentration, and the lines perpendicular to these shells represent the pathways of carbon dioxide molecules diffusing down the concentration gradient. If the dish (Fig. 9(a)) is replaced by one of half the diameter (Fig. 9(b)), the concentration shells form in such a way that the curved surface corresponding to a given concentration, e.g., C in Fig. 9(a) and (b), will be at half the distance from the surface of the dish which has half the diameter, that is to say the concentration gradient, on which diffusive flux depends, has been doubled.

Returning to Fig. 8, 8(b) shows the planes of equal concentration in this system, if the air is still. C_m is unknown, but eqns (1) and (2) both describe the same diffusive flux, though in different sections of the total pathway. They may be combined algebraically to eliminate C_m and give

$$F = \frac{k'(C_a - C_s)}{L + 0.392\,D} \, g/cm^2/s \qquad (3)$$

In general we now see that the diffusive flux through resistances in series is directly proportional to the difference in concentration between the ends of the whole system, and inversely proportional to the sum of the resistances, which in Fig. 8(b) are the length of the tube and a diffusion shell at its mouth whose length is effectively $(\pi/8)D$.

Because the rate of diffusion through unit area of orifice is inversely proportional to its diameter, Brown and Escombe supposed that a screen pierced with many small pores might offer only a small resistance to diffusion, and they conducted a number of experiments like that illustrated in Fig. 10, in which screens, made from photographic film and perforated in different ways, were placed in the mouth of the tube. Fig. 11, reproduced from their results, shows for instance that with screen 3, although only about 1 per cent of the area was perforated, the rate of diffusion per unit area of tube was 40 per cent of that obtained in a similar tube without an obstruction at its mouth (column 5).

By analogy with eqn (3), but remembering that each pore has a diffusion shell at each end, the resistance to diffusive flux *per unit area of pore* is $l + (2 \times 0.392\,d)$, where l and d are length and diameter of the pore. If unit area of the screen has n pores, each of area a, then the area of pores per unit area of screen is na (which is less than 1) and

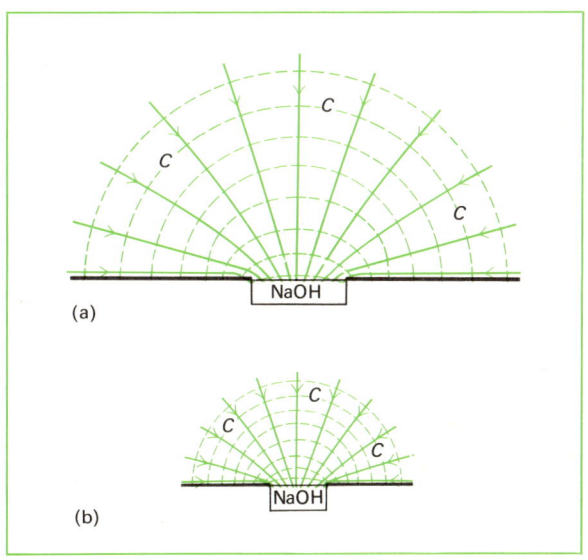

FIG. 9. Diffusion 'shells' over dishes of different diameter; each dish with a broad rim. Dish (a) has twice the diameter of dish (b).

9

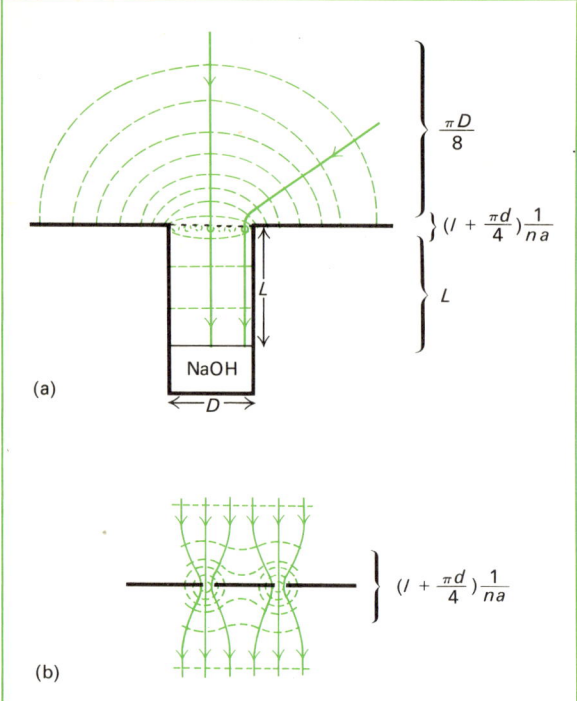

(a)

(b)

FIG. 10. Diffusion of carbon dioxide to sodium hydroxide in a tube whose mouth is covered by a perforated screen. (b) Shows on an enlarged scale the diffusion through two pores in the screen at the mouth of (a).

Screen no.	Pores per cm²	Distance between pores as multiples of pore diameter	Area perforated per cent	Diffusion, as per cent of value from open tube without screen	
				Observed	Calculated
1	100	2·6	11·3	56·1	87·6
2	25	5·3	2·8	51·7	63·7
3	11·11	7·8	1·25	40·6	44·0
4	6·25	10·5	0·7	31·4	30·7
5	4·0	13·1	0·45	20·9	21·9
6	2·77	15·7	0·3	14·0	15·5

FIG. 11. An experiment on diffusion through perforated screens by Brown and Escombe (*Phil. Trans. R. Soc.* B **193**, 1900). The experimental arrangements are shown in Fig. 10. Various screens (1–6) were used, but in each case the thickness of the screen was 0·1 mm and the diameter of the perforations 0·38 mm.

the resistance to diffusive flux *per unit area of screen* is therefore $\dfrac{l + 0.785\,d}{n\,a}$. The sum of all the resistances in the system should then be

$$R = 0.392\,D + \frac{l + 0.785\,d}{n\,a} + L$$

(see Fig. 10) and for screen 3 the quantities on the right-hand side of this equation are equivalent to 1·36, 2·95, and 1·0 cm respectively. Using this formula for diffusive resistance, Brown and Escombe calculated the expected diffusive flux, $F = \dfrac{k'\,(C_a - C_s)}{R}$, for tubes with differently perforated screens. Observed and calculated rates agreed well (Fig. 11, columns 5 and 6) when the pores were widely spaced but the observed rates were less than the calculated rates when the pores were less than 10 diameters apart. They concluded that in the latter conditions, adjacent diffusion shells interfered with each other sufficiently to reduce diffusion.

Consider next the system in Fig. 12, in which sodium hydroxide is replaced with a water surface close below the perforated screen. We now have a model approximating to one side of a transpiring leaf. Water vapour diffuses from the saturated air with high concentration inside the leaf, to unsaturated air outside the leaf, and through the same resistances as would be encountered by carbon dioxide diffusing inwards. Of course the geometry of the stomata and substomatal cavities is much more complex than the model, and most air conditions are not still enough for the formation of a hemispherical diffusion shell on the surface of a large leaf. Rather, a flattened boundary layer is formed (Fig. 6). Nevertheless the resistances are still calculable or measurable. For a leaf surface with 15 000 stomata per cm², each with pore length 10 μm and effective diameter 5 μm, the resistance of the stomatal array will be equivalent to a layer of still air approximately 1·3 cm thick. If the leaf has a mean diameter of 5 cm, the thickness of the boundary layer will be about 1·5 cm in still conditions, decreasing to a few mm in wind.

We now see why the rate of transpiration through an epidermis may be as much as half the evaporation from a wet paper of the same size and shape. Because of the rapidity of diffusion through small pores, the resistance of the epidermis is much less than might be expected, and both the leaf and the paper have a boundary layer resistance. If stomatal and boundary layer resistances are equal the total resistances to diffusion from leaf and paper are in the ratio 2:1. Even out of doors, the leaves of dense vegetation are sufficiently sheltered so that with open stomata and in moderate wind-speeds, the stomatal and external resistances are of similar size. If the stomata close, stomatal resistance increases, and with increasing wind, the external resistance declines (Fig. 13).

To conclude this section, we can express transpiration (E_t) per unit leaf area by the following equation:

$$E_t = \frac{k\,(C_s(T) - C_a)}{l'_s + l'_a} \text{ g water vapour/cm}^2 \text{ leaf area/s} \tag{4}$$

where k is the diffusion coefficient of water vapour in air, $C_s(T)$ is the concentration of water vapour of saturated air within the leaf and is determined by leaf temperature T, C_a is the concentration in the ambient air, and l'_s and l'_a are the effective lengths of the internal and external diffusion paths. (Compare with eqn (1).)

To illustrate the sensitivity of transpiration to the numerator of this expression let us first assume that leaf and air are both at 20°C and the relative humidity of the air is 70%. The concentration of water vapour in the saturated air within the leaf would then be 17·3 g/m³, and of the air outside the boundary layer $17·3 \times \frac{70}{100} = 12·1$ g/m³. The gradient $C_s(T) - C$ would therefore be 5·2 g/m³. Leaves in sunshine are often several degrees warmer than the air around them (see Fig. 14) and since the vapour pressure of saturated air increases rapidly with temperature, leaf temperature is important in determining transpiration rate. For instance if leaf temperature in sunshine rose to 25°C, air conditions remaining the same, the water vapour concentration of saturated air within the leaf would become 23·2 g/m³ and the gradient $C_s(T) - C$, 11·1 g/m³, i.e. the concentration gradient and the transpiration rate would be more than doubled.

Some effects on transpiration of varying l'_s, mainly controlled by the stomata, and l'_a, mainly controlled by wind, have been illustrated in Fig. 13.

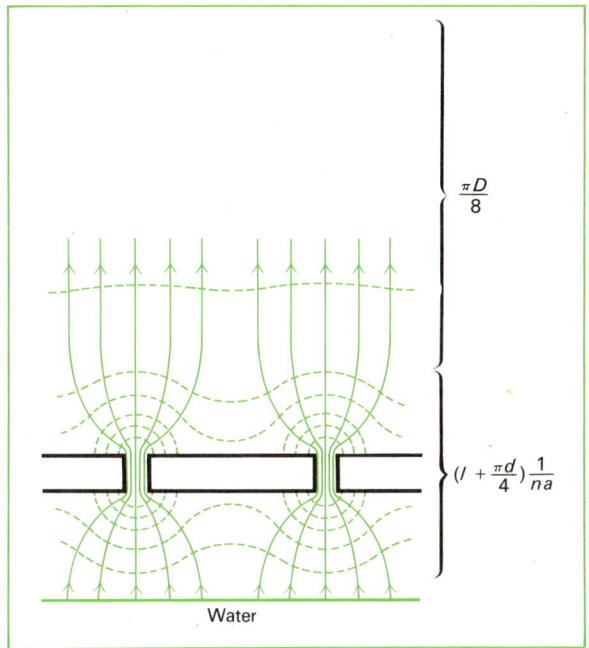

FIG. 12. (left) Diffusion of water vapour from a free water surface through a perforated septum.

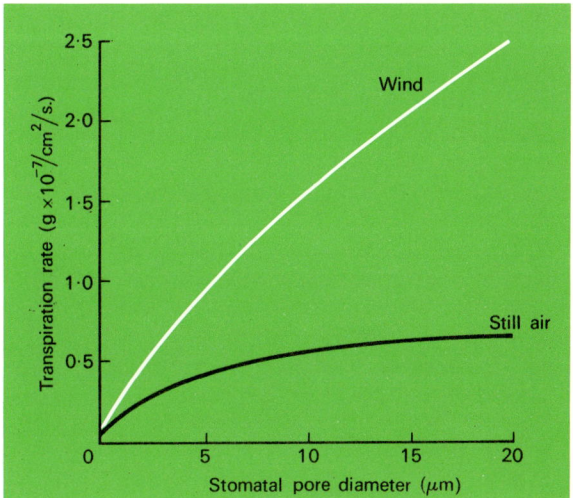

FIG. 13. The response of transpiration to wind and stomatal aperture in *Zebrina pendula*. In still air the external resistance is higher than the stomatal resistance and there is little response to stomatal closure until it is nearly complete. In wind the external resistance is low, transpiration is much higher when the stomata are open, but is responsive to the whole range of stomatal aperture. Note the very low transpiration when the stomata are closed indicating that there is little cuticular transpiration. (After Bange (1953), *Acta bot. neerl.* **2**.)

11

Penman's equation

Most people who have had any experience of drying things, for instance wet clothes, will have an intuitive understanding that evaporation rate increases with temperature and wind, and decreases with atmospheric humidity. Equation (4) gives formal expression to this for $C_s(T)$ is determined by temperature, in this case leaf temperature, C_a is an expression of the humidity of the air, and l'_a varies inversely with wind-speed. The equation, though formally correct, does not go far enough since it does not express the fact that in transpiration energy is required to supply the latent heat of vaporization. In the absence of a supply of energy (heat) the leaf would be rapidly cooled by transpiration and $C_s(T)$ would fall until the gradient $(C_s(T) - C_a)$ became zero. The temperature of a plant in a room is maintained by the air around it, i.e. it obtains its heat from the air and often indirectly from the central-heating system. In nature when the whole surface of the land is covered with transpiring vegetation, the temperature of leaves and air would fall rapidly in the absence of any external source of energy, and plants in nature obtain most of their energy for transpiration by the absorption of solar radiation.

Penman has developed a more complex equation which relates transpiration to the intensity of solar radiation (on which leaf temperature and $C_s(T)$ depend), air temperature, humidity, and wind. This equation also expresses transpiration not as a rate per unit leaf area as in eqn (4) but as a rate per unit land area. Since rainfall on an area is expressed as a depth of water (so many inches or millimetres) i.e. as volume per unit land area, it is convenient for the comparison of gains and losses to express transpiration in the same way. A major advantage of Penman's equation is that all the climatic observations required are obtainable from standard meteorological records.

The physiological significance of transpiration

So far this article has been concerned with an analysis of the physics of transpiration. Many of its physiological effects can however only be appreciated by reference to the flow of water through the plant which transpiration causes. The most important point to grasp is that the plant normally reacts in such a way as to keep uptake and transpiration of water in balance. As the cells of leaves lose water by transpiration they develop a suction (or reduced water potential) and the effect of this is to draw water through the plant from the soil. If uptake of water is less than transpiration, it is clear that there will be a continuing net loss of water from the leaf cells and hence a continuing increase in the suction they exert. But this increase in suction will lead to an increasing rate of uptake until, unless wilting and loss of turgor supervene, uptake equals transpiration. At this stage there will be no further net loss of water from the cells of the leaf. Uptake therefore always tends to come to an approximate balance with transpiration, and the higher the transpiration rate, the greater will be the reduction in leaf water content needed to achieve this balance.

The amount of water transpired by a growing plant is at least ten times as much as it requires for the vacuolation of new cells and one hundred times as much as it combines with carbon dioxide in photosynthesis; and the plant may transpire several times its own weight of water in a hot day. What are the effects of this large flow of water through the plant?

In the first place, the inorganic materials absorbed by the roots are transported to the rest of the plant mainly in the transpiration stream in the xylem. However if we examine the supply of nutrients from the soil to the shoot we find that this does not consist simply of the passive flow of a solution. The soil solution is usually very dilute and the supply of many nutrients depends first of all on the movement into solution of sparingly soluble substances, or the slow release, e.g. of ammonium salts and nitrates, by the microbiological decomposition of organic matter. Once released, these substances may move to the root both by the mass flow of water resulting from water uptake, and by molecular diffusion, and present evidence is that except where roots are very sparse, diffusion is sufficient to account for most of the movement. Uptake of water and solutes by the root are largely independent processes. The uptake of solutes is often from low external concentrations to higher concentration within the plant, and depends on energy-requiring processes in the root cells. Once in the xylem, the distance between root and shoot is such that molecular diffusion would be inadequate to supply the shoot with the nutrients which it needs and the transpiration stream is a convenient method of transport. It is probable however that the rate-limiting processes in nutrition are those of release into the soil solution and movement across living barriers in the root and that the rate of transpiration is much in excess of that required for transport through the plant.

Secondly transpiration plays a regulatory role in leaf temperature. Leaves exposed to bright sunshine absorb radiation and may become several degrees warmer than the air around them. Fig. 14 shows this for a transpiring leaf, and also shows that the temperature was even higher in an adjacent leaf in which transpiration was prevented by the application of a low melting-point wax which did not materially change the absorption of radiation. The utilization of energy in evaporating water has reduced the temperature of the transpiring below the non-transpiring leaf. Transpiration is only one of a number of processes that regulate the temperature of leaves. They also lose heat by radiation and, especially in wind, by convection.

Having noted the possibly beneficial role of transpiration in transport and temperature regulation, we must however recognize that the magnitude of transpiration is a major hazard in the ecology of most terrestrial plants.

It has been noted above that as plants transpire and the water content of their cells is lowered, so they develop a suction (or reduced water potential) which enables them to withdraw water from the

soil. So long as the soil is fairly moist, a fairly small reduction of plant water content develops sufficient suction to keep uptake and transpiration in balance, although in hot climates plants may wilt at midday even when soil moisture is favourable. In the absence of rainfall, however, transpiration rapidly depletes soil water. As the soil dries it yields water less readily and the water content of the plant cells has to be reduced to a lower level before the suction developed is sufficient to keep uptake and transpiration in balance. Ultimately the soil may become so dry that water can no longer be obtained from it until the supply is renewed. The reduction of cell water content, which is therefore related to transpiration and soil water content, reduces the rate of many physiological processes and leads to stomatal closure which, although it reduces transpiration, also reduces the diffusion of carbon dioxide into the leaf.

Plants rooted to about 1 m depth will have the equivalent of about 150 mm depth of water available to them in many soils. With a transpiration rate in the temperate zone summer equivalent to 3–4 mm per day, the available water will be half used, and growth appreciably reduced after less than a month without rain. On shallow soils, or with the higher transpiration rates that are found in subtropical regions, available water in the soil may be quite rapidly exhausted in periods without rain. Standing between an unsaturated atmosphere and a soil of limited storage capacity, the transpiring plant is often between the devil and an insufficiently deep sea. The numerous adaptations shown by xerophytes in this precarious situation are the subject of another contribution to this series (No. 39, *Xerophytes*, by S. R. J. Woodell).

FIG. 14. Temperatures (measured with fine thermocouples) of two adjacent leaves on a plant of *Chenopodium album*, one of which has been coated with a low melting-point wax to prevent transpiration. The difference between leaf temperature and air temperature is plotted. (Data at hourly intervals have been abstracted from the continuous records of Wallace and Clum (1938), *Am. J. Bot.* **25**.)

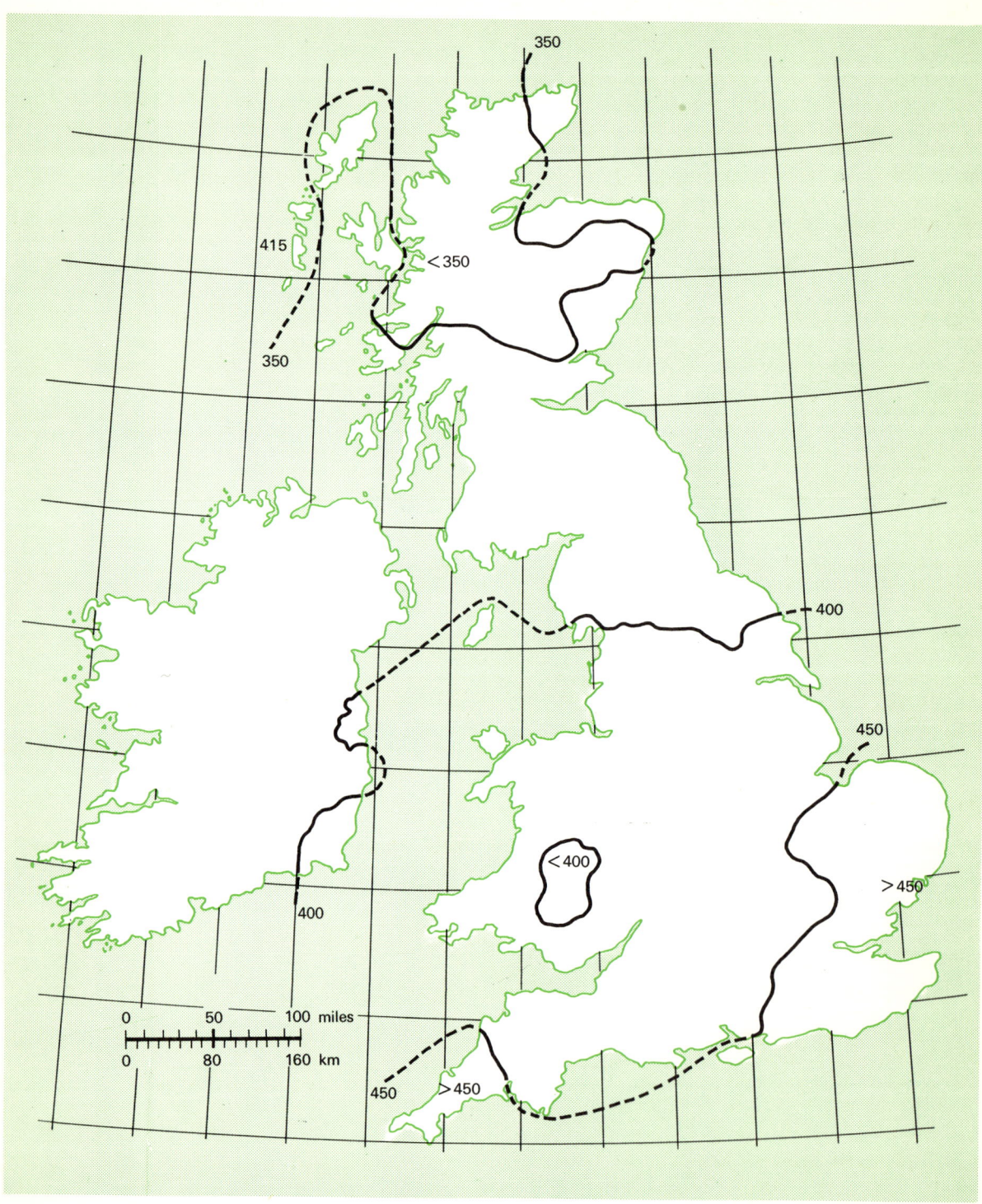

FIG. 15. Average potential transpiration in mm, totalled for the months April–September, in the British Isles. (Simplified from a map given in *Tech. Bull. Minist. Agric. Fish. Fd.* **16**.)

Transpiration and practical affairs

Because the depletion of soil water by transpiration reduces the growth of crops, there are many parts of the world where agriculture is possible only if water is available for irrigation. Even in England rainfall is often insufficient to maintain an optimum level of soil water in the summer months, and crop yields are increased by irrigation. This means that if rainfall is less than transpiration, the difference has to be supplied by irrigation, at least at those stages of growth when crops are particularly sensitive to drought. In order to calculate the amount of water to be given, estimates of transpiration are therefore required. Experiments have shown that the stomatal resistances of different kinds of crop are not very different and that their transpiration rates are sufficiently similar to be calculable from Penman's formula using the standard observations of the Meteorological Office. Fig. 15 shows the calculated depth of water transpired by crops during the summer months of an average year in various parts of the British Isles. There is not much transpiration in winter. Similar figures are available for individual months and a service of the Meteorological Office is available to farmers through which, at the end of each month, they can obtain an estimate of transpiration and correct their irrigation programme for any excess of transpiration over the average figure for the month on which their planned irrigation was calculated.

Assuming an average rainfall of about 300 mm in the summer months in south-east England and (from Fig. 15) a potential transpiration of about 450 mm, it will be seen that 150 mm of irrigation is theoretically required. It will of course only be given to those crops, mostly market-garden crops and early potatoes, in which the increased yield is worth more than the cost of irrigation.

If Fig. 15 is compared with a similar map of rainfall it will be realized that transpiration, along with smaller sources of evaporation, takes a large toll on rainfall and must materially affect the flow of rivers. Measurements or estimates of transpiration are therefore as important to water engineers as measurements of rainfall, to help them to understand and predict the quantity of water that will be available in river flow at different times of the year. In a society whose needs for water for industrial and domestic use are constantly increasing, accurate budgets of available water are extremely important. Estimates of transpiration over a whole river catchment are more difficult than for an irrigated crop, partly because one cannot assume that vegetation types so diverse as arable crops, grassland, heath, and forest will respond in the same way to climatic factors, and also because differences in soil type, depth, and moisture content play a greater part in controlling transpiration than they do in irrigated agriculture. This complex situation is nevertheless capable of both experimental and theoretical analysis. Approximate solutions to many of the problems are already available, and in the United Kingdom a considerable amount of research in this field is carried on by organizations like the Natural Environment Research Council (see Fig. 3) and some universities.

FURTHER READING

General

KOZLOWSKI, T. T. (1964). *Water metabolism in plants.* Harper and Row, New York.

RUTTER, A. J. (1967). Evaporation in forests. *Endeavour* **26**, 39–43.

SUTCLIFFE, J. F. (1968). *Plants and water.* Arnold, London.

For reference

GATES, D. M. (1965). *Energy exchange in the biosphere.* Harper and Row, New York.

KRAMER, P. J. (1969). *Plant and soil water relationships.* McGraw-Hill, New York.

MEIDNER, M. and MANSFIELD, T. A. (1968). *The physiology of stomata.* McGraw-Hill, London.

MINISTRY OF AGRICULTURE, FISHERIES, and FOOD. (1967). Potential transpiration. *Tech. Bull. Minist. Agric. Fish. Fd.* **16**.

PENMAN, H. L. (1963). Vegetation and hydrology. *Tech. Commun., Commonw. Bur. Soils* **53**.

See these other titles in the Oxford Biology Readers series:

9. *Photosynthesis.* C. P. Whittingham.
12. *Mycorrhiza.* J. L. Harley.
15. *Phloem.* F. B. P. Wooding.
37. *Stomata.* O. V. S. Heath.
39. *Xerophytes.* S. R. J. Woodell.

24